NOVICES GUIDE TO COWPEA FARMING

From Seed To Harvest, The Beginners Manual To Cultivating, Achieving Success And Thriving In Cowpea Farming

CHERYL VALERIE

Copyright © 2024 by CHERYL VALERIE

All rights reserved. Except for brief quotations embodied in critical reviews and certain other noncommercial uses permitted by copyright law, no part of this publication may be reproduced, distributed, or transmitted in any form or by any means, Including photocopying, recording, or other electronic or mechanical methods, without the prior written permission of the publisher.

Disclaimer:

This book is a work of fiction/non-fiction, and any resemblance to actual persons, living or dead, or actual events is purely coincidental. The views expressed in this book are solely those of the author and do not necessarily reflect the views of any organization, company, or individual.

The author would like to clarify that they are not in any endorsement deal with any organization, company, or individual mentioned in this book. Any references to products, services, or entities are made for literary or informational purposes and do not constitute an endorsement or promotion.

Readers are advised to consider this work as a creative endeavor, and the author disclaims any responsibility for consequences arising from any actions or decisions based on the content of this book.

Any opinions expressed within are the author's own and should not be construed as professional advice

Contents

CHAPTER ONE 11
  SELECTING THE APPROPRIATE KIND 11
    Planting And Preparing The Land 11
    Prospects For The Market And Financial Sustainability 15
    Participation In The Community And Exchange Of Knowledge 17

CHAPTER TWO 19
  CHOOSING THE APPROPRIATE COWPEA 19
    Well-liked Cowpea Types 19
    Considerations For Variety Selection 20

CHAPTER THREE 23
  GETTING THE SOIL READY FOR COWPEA FARMING 23
    Conditions For Soil: 23
    How To Prepare Soil Correctly: 24
    Leveling The Soil: 25

CHAPTER FOUR 27
  TECHNIQUES FOR PLANTING TO ENSURE IDEAL GROWTH 27

    Selecting The Ideal Type Of Cowpea: 27

    Planting Distance And Alignment: 28

    The Process Of Fertilization 29

CHAPTER FIVE 31

  IDEAL SEASONS FOR PLANTING 31

    Recognizing Seasonal Differences 31

    Guidelines For Depth And Spacing 33

    Changing The Spacing To Account For Climbing Types: 34

CHAPTER SIX 37

  ESSENTIAL METHODS FOR CROP CARE 37

    Tips For Irrigation And Watering 37

    Techniques For Weed Control 39

    Guidelines For Fertilization 40

CHAPTER SEVEN 43

  COMPREHENDING THE STAGES OF COWPEA GROWTH 43

    From Germination To The Stage Of Seedlings 43

    Blooming And Pod Creation 45

    Harvesting And Ripening 46

CHAPTER EIGHT 49
TRACKING DEVELOPMENT AND GROWTH ... 49
- Monitoring Of Soil 49
- Water Resources Management 50
- Harvesting And Handling After Harvest 52

CHAPTER NINE 57
COMMERCIAL METHODS FOR COWPEA FARMERS 57
- Finding The Right Target Markets 57
- Developing Connections With Purchasers .. 58

CHAPTER TEN 63
FAQs AND REGULAR QUESTIONS 63
- Selecting The Appropriate Variety: 63
- Overcoming Typical Obstacles In Cowpea Farming 66
- Commonly Asked Questions 69

The thorough manual "Cowpea Farming" explores the nuances of cowpea cultivation and offers a plethora of information to both beginning and seasoned farmers. The book begins with a perceptive introduction that establishes the foundation for comprehending the fundamentals of cowpea farming as well as its wider agricultural relevance.

The overview of cowpeas provides readers with a thorough explanation of the traits, growth patterns, and several advantages of cultivating the plant, making it a useful place to start.

Next, the advantages and significance of cowpea growing are emphasized, with a focus on its contribution to sustainable agriculture, nutritional value, and financial consequences for farmers.

A key component of effective cowpea cultivation is variety selection, which is handled in this book by providing a choice of well-liked cowpea varieties. Considerations for variety selection are examined to help farmers make decisions that are well-informed and suited to their unique requirements and local environment.

Next, the guide delves into the critical processes of soil preparation, clarifying the needs of the soil and offering a thorough plan for the best possible cultivation. The planting techniques section provides a strong foundation for cowpea growth by outlining the ideal planting seasons as well as suggestions for exact spacing and depth.

The book then moves on to crucial crop care procedures, covering important topics like fertilizer recommendations, weed control techniques, and irrigation and watering advice. It also provides insightful information on controlling illnesses and pests, including methods for detection, avoidance, and management.

Farmers must comprehend the various stages of cowpea growth, and this book covers all of the details, from germination to harvest, including monitoring growth and development, harvesting techniques, and post-harvest handling.

It places a strong emphasis on mature crop indicators, harvesting practices, and appropriate handling, storing, and processing techniques.

The final sections of the guide expand its scope to include the market by providing cowpea farmers with ideas on how to pinpoint target markets, cultivate connections with buyers, and use efficient pricing and selling techniques.

Common questions are addressed by the addition of FAQs and a troubleshooting guide, which offers farmers a useful tool for overcoming obstacles in the ever-changing area of cowpea farming. This book positions itself as an essential resource for anyone hoping to succeed in cowpea farming, elevating the conversation surrounding this practice.

CHAPTER ONE

SELECTING THE APPROPRIATE KIND

One of the most important things to do when starting a cowpea farm is to choose the best type for your intended use and geographic area. There are many varieties of cowpeas, and each is suitable for a certain type of soil and climate. Certain types exhibit greater resilience against pests and illnesses, whereas others flourish in particular temperature ranges. To find the best cowpea variety for your location, do some study and speak with local agricultural experts before starting your cowpea farm.

Planting And Preparing The Land

Effective preparation of the ground is essential for a successful cowpea harvest. Start by removing any rocks, weeds, and debris from the chosen area of ground. Cowpeas need well-drained soil to flourish, so make sure your drainage is adequate to avoid waterlogging. For cowpeas, the optimal pH range is 6.0 to 7.5. To improve soil fertility, plow the area down to a depth of 6 to 8 inches and add organic matter. When the ground is ready, evenly distribute the cowpea seeds by planting them at the suggested spacing. For best results, plant in the right season, which is usually after the last frost.

Techniques for Watering and Irrigation

Cowpeas need steady moisture levels to develop well and produce large amounts. Watering plants enough is crucial, especially while they are in the flowering and pod-setting stages. Using effective irrigation methods, like furrow or drip irrigation, helps preserve water and guarantees that the plants get plenty of it. But be careful not to overwater; too much moisture might cause fungus-related illnesses. Maintaining a healthy cowpea crop requires regular monitoring of soil moisture levels and changing irrigation schedules accordingly.

Controlling Nutrients and Fertilization

Like all crops, cowpeas gain from appropriate nutrition management. Test the soil for nutrients in your field and apply fertilizers based on the results. For best growth, cowpeas typically need nitrogen, phosphorus, and potassium. To improve soil fertility, organic fertilizers can also be added, such as compost or well-rotted manure. Follow the prescribed rates and time for fertilizer application; overuse can cause nutrient imbalances and harm the crop.

Control of Diseases and Pests

Proactive management of pests and diseases is essential to a successful cowpea yield. Check the plants frequently for indications of pests such as pod borers, thrips, or aphids. Use integrated pest management techniques, such as using appropriate insecticides when needed and natural predators when appropriate. Crop rotation is another way to lower the incidence of soil-borne illnesses. A healthy cowpea crop is largely dependent on early detection and timely action.

Harvesting and Handling After Harvest

When it comes to collecting cowpeas, timing is everything. To guarantee the best flavor and nutritional content, the pods should be gathered at the appropriate maturation level. Harvesting at the wrong time of year can lower the quality. Handle the cowpeas gently after harvesting to avoid damaging the pods. After harvesting, cowpeas should be properly dried and stored in a cool, dry location. Good post-harvest handling techniques guarantee that the cowpeas keep their quality and are prepared for market or storage.

Prospects For The Market And Financial Sustainability

In addition to providing for the food needs of the community, cowpea farming offers economic potential.

Examine local markets, supermarkets, and even export opportunities as possible markets for your cowpea produce. Comprehending client inclinations, price tactics, and industry trends can enhance the financial sustainability of your cowpea farming endeavor.

To stay up to date on market dynamics and make decisions that will increase the profitability of your cowpea farming business, think about establishing a network with regional agricultural extension agencies and market specialists.

Environmental Aspects and Sustainable Practices

Long-term cowpea farming success depends on adopting sustainable farming methods. Using organic agricultural techniques, like crop rotation and cover crops, improves soil health and lessens the need for artificial inputs. Environmental conservation is also aided by the use of natural pest control techniques and agroforestry activities.

Cowpea growers can maintain the health of their farms and improve the environment as a whole by implementing sustainable methods.

Participation In The Community And Exchange Of Knowledge

Good things can come from interacting with the neighborhood and imparting knowledge about cowpea farming.

Work together to plan workshops and training sessions with local farmer's organizations, agricultural extension services, and educational institutions.

By exchanging insights, best practices, and the advantages of cowpea farming, one might inspire others to pursue this business. Participation in the community also makes it easier for insightful information to be shared and builds a network of support for cowpea growers. The success of cowpea cultivation can benefit the entire community if a culture of knowledge-sharing is promoted.

CHAPTER TWO

CHOOSING THE APPROPRIATE COWPEA

When the correct variety is selected, according to your goals and the unique conditions of your area, cowpea farming can be a lucrative endeavor. Choosing the right cowpea variety is essential to maximizing yields and guaranteeing a good harvest.

Well-liked Cowpea Types

Some types of cowpeas have become more well-known in the field of cowpea farming because of their environmental adaptation and performance. Among these is the IT89KD-288 variety, which is well known for its resistance to bacterial blight and for growing well in semi-arid and arid environments. The California Blackeye is another alternative worth considering. It is well-known for its great yield potential and exceptional resistance to common pests.

It's important to examine your farm's unique agro-climatic conditions while choosing cultivars. For example, The Big Buff grows well in warm climates and continues to grow strong even during intense heat waves. Conversely, the Vita 4 type is a great option for areas that are prone to drought because of its remarkable water-use efficiency.

Considerations For Variety Selection

Climate Suitability: The effectiveness of your cowpea farming is greatly influenced by the climate in your area. Choosing a type that will flourish in your particular location will be made easier if you are aware of the temperature, rainfall patterns, and any potential pest pressures.

Disease Resistance: Choosing cultivars that are naturally resistant to common illnesses is a wise move. For example, the IT97K-499-35 demonstrates resistance to the common hazards of cowpea crops, Striga gesnerioides, and Alectra vogelii.

Growth Duration: Take into account the cowpea variety's growth duration and match it to the current growing season in your area. Short-duration cultivars, such as the Oloyin variety, work well in climates with short growing seasons.

Yield Potential: To maximize production, a variety's yield potential must be evaluated. Aiming for maximum yield, farmers often use Blackeye 46 and Blackeye 47 varieties because of their reputation for high yields.

Market Demand: Keep yourself updated on the types of cowpeas that are in demand. Depending on the financial prospects of your agricultural operation, you may choose to plant a particular variety over others based on which are more popular in regional or global markets.

You may create a strong basis for a profitable and successful cowpea farming endeavor by carefully weighing these aspects and choosing a cowpea variety that fits the unique conditions and objectives of your farm.

CHAPTER THREE

GETTING THE SOIL READY FOR COWPEA FARMING

Conditions For Soil:

Cowpeas grow best on well-drained soil that has a pH between 6.0 and 7.5, which is slightly acidic to neutral. Given that cowpeas are known to adore sunlight, selecting a location with adequate exposure is crucial to a successful crop. Loamy soil is great because it offers a harmonious combination of sand, silt, and clay. This makes it possible for the right drainage, aeration, and nutrient retention.

It is helpful to perform a soil test before beginning the soil preparation process. You can use this analysis to assess the soil's nutrient levels and make well-informed fertilization recommendations. Cowpeas demand a modest level of fertility; a soil test will help you choose how much organic or synthetic fertilizer to use.

How To Prepare Soil Correctly:

Clearing and Cleaning the Site: Start by clearing the chosen area of any trash, weeds, or undesirable plants. This stage guarantees a spotless area for the best possible cowpea growth. To efficiently clear the soil and pull weeds, you can use a shovel or hoe.

After the area has been cleared, till or plow the soil down to a depth of 6 to 8 inches. This procedure promotes improved root penetration, increases aeration, and loosens up compacted soil. Utilize a hand tiller, plow, or tractor, based on how big your cultivating area is.

Adding Organic Matter: Adding organic matter, such as manure or well-rotted compost, can improve soil fertility. Using a rake or tiller, cover the plowed area with a layer of compost and work it into the soil. This procedure encourages a thriving microbial community in addition to enriching the soil with vital nutrients.

Leveling The Soil:

To guarantee consistent planting and irrigation, level the prepared soil surface. A uniform distribution of nutrients is facilitated and localized waterlogging is avoided on a flat surface. To prepare a surface for planting that is level and smooth, use a rake.

Building Raised Beds: If your area has inadequate drainage, you might want to think about building raised beds.

Raised beds provide you with better control over the quality of the soil, improved drainage, and a lower chance of soggy roots. If you decide to use raised beds, make sure they are tall enough and wide enough to allow for the proper growth of cowpeas.

Mulching (Optional): Mulching helps maintain soil moisture, prevent weed growth, and control soil temperature.

Spread a layer of hay or straw mulch, or organic mulch, around the cowpea plants. Mulching also reduces soil erosion and helps control weeds.

These procedures will help you create the perfect environment for cowpea farming. A successful farming endeavor, increased yield, and robust plant development are all facilitated by proper soil preparation.

CHAPTER FOUR

TECHNIQUES FOR PLANTING TO ENSURE IDEAL GROWTH

Selecting The Ideal Type Of Cowpea:

A successful crop of cowpeas depends on choosing the right type. The growth patterns, disease, and insect resistance, and climatic adaptation of different varieties vary. Before planting, do your homework and select a variety based on your agricultural objectives and local conditions.

Choosing and Caring for Seeds:

To guarantee a nutritious crop, plant seeds of the highest caliber. To prevent soil-borne illnesses, apply fungicides or insecticides to the seeds before sowing. This pre-treatment increases germination rates and aids in the early establishment of a healthy crop.

Getting the Soil Ready:

Sand loam soils that range in pH from slightly acidic to neutral are ideal for cowpea growth; harrow and plow the soil to create a fine seedbed. To increase the fertility and structure of the soil, add organic materials, such as compost or well-rotted manure.

Planting Distance And Alignment:

The ideal planting depth for cowpea seeds is one to two inches. Plants need to be spaced appropriately apart to grow and develop properly. As a general rule, 4 to 8 inches should be left between plants and 24 to 36 inches between rows. However, pay attention to the spacing guidelines specific to your variety.

Watering Methods:

Water your plants regularly, especially in the germination and blossoming stages. Although cowpeas can withstand some drought, consistent irrigation is still necessary to ensure optimal pod development and output. Minimize the amount of moisture on the foliage by applying water directly to the base of the plant using soaker hoses or drip irrigation.

Weed Management

Put in place efficient weed management strategies to stop competition for sunshine and nutrients. Mulching the area around the plants helps keep the soil temperature more consistent, reduce weed growth, and preserve moisture.

The Process Of Fertilization

Before planting, apply a fertilizer that is balanced and contains potassium, phosphorus, and nitrogen. During the growing season, more fertilizer can be required, particularly if your soil is deficient in important nutrients. To ascertain the precise nutrient needs of your cowpea crop, examine the soil.

Systems of Support:

Support systems might be advantageous for some cowpea varieties, particularly climbing or vining variants. Put in stakes or trellises to give the plants structure and shield them from harm as they grow.

CHAPTER FIVE

IDEAL SEASONS FOR PLANTING

Recognizing Seasonal Differences

Because cowpeas are sensitive to the environment, planting during the appropriate season is essential. The best time to plant varies according to your climate and geographic location. Cowpeas should be planted after the last frost has passed because they generally like warm weather.

Planting in the spring:

Cowpeas are best planted in the spring in areas with mild winters. This enables the crop to grow vigorously by utilizing the warm temperatures and extended sunshine hours. Before planting seeds, make sure the soil has warmed to at least 60°F.

Planting in the summer:

Summer planting can be successful in warmer climates. The ideal temperature range for cowpeas is 70°F to 95°F. During the hot summer months, pay attention to the availability of water since cowpeas need steady hydration to grow to their full potential.

Planting in the Fall:

It is possible to plant in the fall in regions with longer growing seasons. This enables cowpeas to reach maturity before the arrival of frost. It is ensured that the crop receives enough warmth for germination and early growth by planting in late summer or early fall.

Planting in the Winter:

Cowpeas are susceptible to frost, so avoid winter planting in areas where temperatures drop below freezing. Winter planting may cause reduced growth and poor germination. To optimum output, planting schedules must be coordinated with the local climate.

Guidelines For Depth And Spacing

Plant Spacing Calculation:

For cowpea plants to develop and yield as best they can, proper spacing between them is essential. Pay attention to the spacing recommendations given for the cultivar you have selected. The usual spacing between plants is 4 to 8 inches, while the spacing between rows is 24 to 36 inches. Good air circulation is made possible by enough space, which lowers the risk of illness.

Taking Planting Depth Into Account

Sow cowpea seeds in well-prepared soil, letting them sink one to two inches. While planting too deep may cause germination to be delayed, planting too shallow exposes seeds to drying out. To ensure consistent emergence and germination, make sure the planting depth is uniform.

Effective Row Spacing:

To maximize space usage and facilitate farm operations, row spacing is equally vital. Row spacing should be kept between 24 and 36 inches, per the guidelines for the type you have selected. Ease of cultivation, harvesting, and pest control are made possible by ample row spacing.

Changing The Spacing To Account For Climbing Types:

For cowpea kinds that climb or vine, give them more space and provide support in the form of trellises or stakes. Due to the spreading nature of these kinds, adjust the spacing to avoid crowding. In dense canopies, adequate support structures guarantee appropriate air circulation and reduce the danger of disease.

Keeping an eye on and modifying spacing during growth:

Keep an eye on the growth of your plants and alter the spacing as needed. Plants in densely populated places can become healthier when there is less competition for sunshine and nutrients.

To further improve total agricultural productivity and maximize area use, take into account intercropping techniques.

CHAPTER SIX

ESSENTIAL METHODS FOR CROP CARE

Tips For Irrigation And Watering

To achieve maximum development and productivity, cowpea farming requires careful irrigation and watering.

Having enough moisture is crucial, especially when the plant is germinating and blossoming. Establishing a regular watering schedule is essential to giving the plants the appropriate amount of water at the appropriate time.

As cowpeas like well-drained soil, it's critical to prevent overwatering, which can cause root rot.

For cowpea farming, drip irrigation systems are highly recommended since they effectively direct water to the root zone of the plant, reducing water waste.

Farmers should routinely check the moisture content of their soil and modify the frequency of irrigation according to the weather. It's critical to water more frequently during dry spells to keep the soil continuously hydrated.

Water conservation can also be significantly aided by mulching the area surrounding the cowpea plants.

An organic mulch layer controls soil temperature, inhibits weed growth, and helps keep soil moisture in the soil.

As a result, less frequent watering is required, which improves the overall sustainability and efficiency of the irrigation process.

Techniques For Weed Control

Maximizing output and encouraging cowpea growth requires efficient weed control. Cowpea plants are hampered in their growth by weed competition for nutrients, water, and sunlight. Weeding by hand is an essential technique, particularly in the early phases of cowpea development.

To stop weeds from becoming a dominant presence, farmers should manually remove them from their fields regularly.

Weed growth can also be inhibited by using cover crops or intercropping with compatible plants. Cover crops are positioned strategically to create a natural barrier that prevents weed germination and growth. Another successful weed-control strategy is mulching, which not only retains moisture but also inhibits weed seed germination and blocks sunlight, therefore slowing down weed development.

Pre-emergence herbicides can be used sparingly in larger-scale operations, adhering to suggested criteria to prevent harm to the cowpea plants. To avoid harming the crop, it is crucial to use herbicides carefully, making sure they are suitable for cowpeas and sprayed at the appropriate period of development.

Guidelines For Fertilization

Strong cowpea yields are largely dependent on proper fertilization. Choosing the right fertilizer application method requires knowledge of the nutrients needed at each stage of growth. Generally speaking, cowpeas are nitrogen-fixing legumes, meaning they work in symbiosis with nitrogen-fixing bacteria in their root nodules to obtain nitrogen from the atmosphere. For optimum growth, they still need enough potassium, phosphorus, and other minerals.

It is best to test the soil before planting to ascertain current nutrient levels and deficits. Farmers can create a targeted fertilization plan based on the findings. The cowpea plants have a nutrient-rich base when the soil is amended with well-decomposed organic matter before planting.

Supplemental fertilization may be required during the growing season, particularly if shortages in certain nutrients are found. Because nitrogen-fixing bacteria are sensitive to elevated nitrogen levels in the soil, caution should be exercised when applying fertilizers containing nitrogen. Appropriate amounts of potassium and phosphorus in balanced fertilizers can promote the general health and growth of plants.

Successful cowpea cultivation requires a comprehensive approach to crop care techniques, such as appropriate irrigation and watering, efficient weed control measures, and targeted fertilization. Farmers can contribute to a successful agricultural operation by enhancing the production and resilience of their cowpea crops through the use of these instructions.

CHAPTER SEVEN

COMPREHENDING THE STAGES OF COWPEA GROWTH

From Germination To The Stage Of Seedlings

The critical germination stage marks the start of the cowpea farming experience. When the ideal conditions of soil moisture, temperature, and air pressure are met, seeds begin to emerge from their dormant state. Cowpea seeds usually appear as small seedlings that break through the soil's surface 5 to 10 days after they are planted. It is crucial to maintain a constant moisture content in the soil during this time to facilitate the delicate emergence of the young plants.

It's critical to monitor possible problems like pests and illnesses as the seedlings emerge. Without sacrificing the general health of your cowpea crop, you can protect the susceptible seedlings by using natural therapies or eco-friendly pesticides.

Growth of Vegetation

The emphasis switches to vegetative growth when the cowpea seedlings have taken root. The growth of leaves and stems characterizes this stage. Getting enough sunshine and having a healthy supply of nutrients is critical at this time. It's well known that cowpeas can fix nitrogen in the soil, which lessens the requirement for excessive fertilizer. To make sure the plant's nutritional needs are satisfied and to encourage robust foliage and strong stems, it is advised to conduct regular soil tests.

At this point, proper plant spacing is essential for optimum air circulation and to reduce the risk of fungal diseases. To ensure prompt interventions and promote a robust and healthy vegetative phase, it is important to regularly monitor the plants for indications of nutritional deficits or pest infestations.

Blooming And Pod Creation

In cowpea cultivation, the shift to the blooming and pod-forming stage is crucial. Delicate flowers appear during this period, and if they are successfully pollinated, they develop into the distinctive pods that hold the cowpea seeds. During this time, bees and other pollinators are essential because they increase the likelihood of successful fertilization.

Flower abortion must be avoided by keeping the soil at its ideal moisture content and preventing over-irrigation while the flowers are blooming. Furthermore, supporting the emerging plants helps avoid harming the developing pods, particularly when producing vining kinds. To prevent problems from getting worse and to safeguard the developing cowpea pods, it is crucial to regularly check the crop for indications of pests or illnesses.

Harvesting And Ripening

The ripening and harvest stage is the last phase of the cowpea growth cycle. The pods become solid to the touch and change color as they ripen, signifying that the seeds within are ready to be harvested.

Care should be taken when harvesting to prevent harming the plants and jeopardizing subsequent crops. Timing is critical depending on the intended purpose, such as drying or fresh eating.

To avoid post-harvest losses, the harvested cowpeas must be thoroughly cleaned and dried. A cool, dry atmosphere is one of the ideal storage settings to preserve the quality of the harvested crop. Farmers may maximize cowpea yields and contribute to the overall success of their agricultural endeavors by being aware of and actively regulating each stage of the growth process.

CHAPTER EIGHT

TRACKING DEVELOPMENT AND GROWTH

Monitoring Of Soil

Monitoring soil conditions regularly is essential to effective cowpea production. Before planting, testing the soil yields important information about pH, nutrient levels, and general health of the soil.

These findings enable farmers to apply fertilizer with knowledge and ensure the best possible growth for cowpea plants.

Frequent inspections throughout the growth phases aid in identifying any inadequacies or imbalances, enabling prompt corrections and averting possible crop problems.

Disease and Pest Monitoring

It is crucial to keep a close eye out for diseases and pests during the cowpea growth cycle. Early threat identification allows for timely intervention, reducing the impact on crop health.

Effective methods for managing pests include using organic insecticides, rotating crops, and utilizing natural predators. It is essential to regularly inspect the plants, paying special attention to the undersides of the leaves and emerging pods, to identify and correct any problems before they get worse.

Water Resources Management

An effective water management strategy is essential for cowpea production. It is crucial to keep an eye on soil moisture levels and modify watering techniques based on the stage of growth. While underwatering can impede plant development, overwatering can cause root rot and other water-related problems. Throughout the cowpea cultivation process, optimal water management can be achieved by installing a dependable irrigation system and modifying the frequency and amount of water applied based on plant requirements and weather conditions.

Management of Nutrients

Even though they are renowned for their capacity to fix nitrogen in the soil, cowpeas nevertheless gain from a well-balanced diet. Keeping an eye on the levels of nutrients during the growth phases guarantees that the plants get the nutrition they need for the best possible development. Frequent foliar analyses and soil tests yield useful information for modifying fertilizer applications. Using sustainable techniques improves nutrient availability and supports the general health of the cowpea crop. Examples of these activities include adding cover crops and organic matter to the soil.

Through careful observation of these critical elements during the cowpea plant's growth and development, farmers can anticipate problems, make the most use of available resources, and cultivate a robust and fruitful harvest.

Harvesting And Handling After Harvest

Indices of Adulthood:

Determining the best time to harvest cowpeas is essential to guaranteeing a plentiful crop of high-quality food. The pods' changing color is one of the primary indicators that cowpea plants have reached maturity. When cowpea pods reach maturity, they often take on a consistent green hue and harden up into plump pods. Furthermore, the plant may begin to turn yellow in the leaves, which would be an indication that it is fully mature. Because harvesting cowpeas at the proper time is critical to the crop's overall viability, farmers should check their plants frequently to note these visual indications.

Methods of Harvesting:

Selecting the appropriate harvesting method is crucial to minimizing plant damage and optimizing production. For cowpeas, hand harvesting is a popular and efficient technique. This is harvesting the mature pods by hand from the plants. Using sharp harvesting implements, such as knives or shears, is essential to preventing needless strain on the plants. The cowpeas' quality may be impacted by bruising or other damage to the pods, thus caution should be used when collecting them.

In addition, when the plants are less stressed by the heat, it is best to harvest in the morning or late afternoon.

For larger-scale enterprises, mechanical harvesting is an additional alternative. Collecting cowpea pods can be done effectively with the help of specialized machinery like combine harvesters. Nonetheless, it's critical to accurately calibrate the equipment to prevent undue pod damage. Whatever the approach, regular crop monitoring and modifying harvesting methods as necessary are essential to a good harvest.

Appropriate Processing and Storage: Following a fruitful harvest, appropriate processing and storage are essential to preserving the grade and market value of cowpeas. To start, remove the seeds from the pods and other plant material by threshing the harvested pods.

Depending on the size of the operation, either manual or mechanical equipment can be used for threshing.

Cowpea seeds must be thoroughly dried after separation to lower their moisture content. Sun drying is a popular and economical technique, although caution should be used to prevent overexposure to sunlight as this may cause discoloration. The moisture percentage of properly dried cowpeas should be between 10 and 13 percent.

To avoid moisture buildup, store the dried cowpea seeds in containers with good ventilation. It's critical to routinely check seeds that have been stored for evidence of mold or pests. Think about storing the seeds for an extended period in airtight containers to keep the seeds safe from the elements.

Using efficient harvesting methods and post-harvest handling practices is essential to maximizing cowpea production and quality. Farmers can ensure a successful and lucrative cowpea farming operation by observing signals of maturity, using appropriate harvesting techniques, and putting correct storage and processing procedures in place.

CHAPTER NINE

COMMERCIAL METHODS FOR COWPEA FARMERS

Finding The Right Target Markets

Farmers must carefully select and comprehend their target audiences to successfully market cowpeas. It is crucial to do detailed research on market trends, geographical demand, and consumer preferences.

This entails researching the nutritional preferences, age ranges, and socioeconomic status of potential customers.

Additionally, farmers can tailor their marketing strategies to the need for cowpeas in various geographic areas, including both urban and rural ones.

Farmers should modify their language and product presentation to appeal to the particular requirements and preferences of that market segment after identifying their target market. For instance, emphasizing cowpeas' nutritional advantages can be a crucial selling feature if the target market comprises health-conscious people.

Surveys and focus groups with the local community can yield insightful information about customer expectations and aid in the improvement of marketing tactics.

Developing Connections With Purchasers

Building trusting ties with purchasers is essential to cowpea farming's success. Farmers ought to aggressively interact with markets, grocery shops, and even neighborhood eateries as possible customers.

Interacting with buyers and comprehending their needs, entails networking and going to industry events. Establishing a dependable and uniform supply chain is crucial for winning over customers and guaranteeing a constant market for cowpeas.

To increase their market share, farmers can also look into joint ventures with distributors or wholesalers.

Working together with these middlemen can improve cowpeas' market visibility and expedite the distribution process. Keeping lines of communication open, attending to customers' issues, and continuously producing high-quality goods are further ways to establish enduring partnerships that are advantageous to both sides.

Selling and Pricing Strategies

Choosing the appropriate pricing plan is essential to cowpea marketing success. When choosing their prices, farmers should take into account variables including production costs, market demand, and rival prices.

Keeping a competitive advantage in the market and drawing in customers requires providing competitive rates while retaining profitability.

Farmers can increase sales by using a variety of marketing strategies in addition to competitive pricing.

This could entail designing eye-catching packaging, advertising exclusive deals or discounts at busy times, and utilizing digital channels to boost online sales.

To improve the overall customer experience, farmers can also think about adding value to their cowpea products. Some ideas for this include supplying pre-packaged convenience choices or recipe suggestions.

Farmers can increase sales and improve the market impression of their cowpea products by implementing efficient pricing and selling strategies. In the highly competitive agricultural world, sustainable success can be achieved by routinely evaluating market trends and adjusting plans accordingly.

CHAPTER TEN

FAQs AND REGULAR QUESTIONS

Selecting The Appropriate Variety:

Choosing the right variety for your particular conditions is one of the first things to take into account when farming cowpeas. Your chosen variety should be appropriate for your intended use, soil type, and climate. Common variants with distinct qualities are IT89KD-288, IT93K-452-1, and IT86D-1010. Seek advice from extension services or local agricultural specialists for recommendations specific to your area to make an informed decision.

Ideal Timing for Planting:

A great harvest depends on choosing the ideal time to sow cowpeas. It is best to plant cowpeas after the last frost when the earth has warmed up, as they like warm weather. This usually occurs in the late spring or early summer in many places. However, depending on where you are, the precise timing could change. For the best time to sow your cowpea crop, pay attention to past weather data and local climatic patterns.

Fertilization and Soil Preparation:

Adequate soil preparation is essential to the success of cowpea growing. Cowpeas want their soils to have a pH between 6.0 and 7.5 and to be well-drained. Do a soil test before planting to determine pH and nutrient levels, and then apply the appropriate amendments. Add organic matter to the soil to increase fertility and soil structure. In general, a balanced fertilizer with an N-P-K ratio of 14-14-14 or something similar is appropriate for fertilization. The health and yield of your cowpea plants are greatly enhanced by proper soil preparation and fertilizer.

Watering Techniques:

Although cowpeas can withstand some drought, regular and sufficient irrigation is still necessary for the best possible development and output. Make sure there is enough moisture during the flowering and pod-setting phases to support the formation of fruit. On the other hand, steer clear of soggy circumstances because cowpeas can develop root rot in too-wet soil. Using appropriate irrigation techniques, like drip or furrow irrigation, can assist in efficiently managing the distribution of water and averting problems associated with water.

Management of Pests and Diseases:

For a plentiful harvest, safeguarding your cowpea crop against pests and illnesses is essential. Aphids, thrips, and pod borers are common pests, and bacterial blight and powdery mildew are illnesses that can be dangerous. Use integrated pest management (IPM) techniques, such as crop rotation, the use of natural predators, and the sparing application of organic pesticides. In cowpea farming, regular scouting and early identification are essential elements of an efficient pest and disease management program.

Overcoming Typical Obstacles In Cowpea Farming

Strategies for Weed Control:

Cowpea yields can be greatly impacted by weed competition, so it's critical to put in place efficient weed control strategies. Common techniques include hand weeding, mulching, and applying pre-and post-emergence herbicides. To prevent crop damage, it is imperative to carefully choose herbicides based on their compatibility with cowpeas and adhere to prescribed application rates. Throughout the cowpea growing season, weed pressures can be reduced by integrating these techniques.

Identification of Nutrient Deficiencies:

Reduced yields can result from cowpea growth and development being impeded by nutrient shortages. Keep an eye out for any indications of shortages in the plants, such as yellowing leaves or reduced growth. Immediately address nutrient deficits by adding the right soil amendments or fertilizers. Regular soil testing can help you tailor the amount of nutrients to your cowpea crop's needs and make sure it gets the nutrients it needs to produce at its best.

Adapting to Climate Variability:

Variability in the climate affects crop growth patterns and yields, which is a challenge in agriculture.

Consider using resilient farming techniques to adjust to the changing climate. This could entail choosing cowpea cultivars that are sensitive to climate change, modifying the timing of plantings in response to climatic projections, and implementing water-saving technology.

Keeping up with regional climate trends and working with agricultural extension services can give you important information about how to modify your cowpea farming methods in response to the changing climate.

Commonly Asked Questions

Q1: Which cowpea variety is ideal for my area?

A1: Your particular climate, soil type, and intended usage will determine which cowpea variety is best for you. To find the best kinds for your area, speak with extension agencies or local agricultural specialists. Typical variants are IT89KD-288, IT93K-452-1, and IT86D-1010; each has unique qualities appropriate for various circumstances.

Q2: What time of year is best to grow cowpeas?

A2: Warm weather is ideal for cowpea growth. Plant them in late spring or early summer, once the earth has warmed up following the last frost. However, depending on where you live, planting schedules may change. The ideal planting window for your cowpea crop will be determined by keeping an eye on past weather data and local climatic patterns.

Q3: In cowpea farming, how can I recognize and control common pests?

A3: Put into practice integrated pest management (IPM) techniques, such as crop rotation, the use of organic insecticides sparingly, and the utilization of natural predators. Scouting regularly is necessary to spot pests early. Aphids, thrips, and pod borers are common pests. React quickly to save your cowpea crop from pest damage.

Q4: In cowpea farming, what are the most important methods for preparing the soil?

A4: Cowpeas like well-drained soils with a pH between 6.0 and 7.5. To determine nutrient levels and pH, first test the soil. Add organic materials to the soil to increase its fertility and structure. Utilize a balanced fertilizer with an N-P-K ratio of 14-14-14 or comparable, and adhere to the advised fertilization procedures.

Troubleshooting Manual

The issue is yellowing leaves and growth retardation.

Solution: To determine nutritional inadequacies, test the soil. Apply the proper soil amendments or fertilizers to address deficiencies. Keep a close eye on the plants and modify the fertilizing schedule according to their needs.

Issue: A weed invasion is hindering the growth of cowpeas.

The answer is to employ a mix of hand weeding, mulching, and herbicides for both pre-and post-emergence. To avoid causing harm to the crop, choose herbicides that are suitable for cowpeas with care and use them at the prescribed rates.

Issue: Variability in the climate has an impact on cowpea yields

Solution: Choose adaptable cowpea cultivars, modify planting dates by climate projections, and apply water-conserving technology to adapt to changing climate conditions. Keep up with regional climate trends and seek advice from agricultural extension services.

You can confidently manage the difficulties of cowpea cultivation and ensure a fruitful and satisfying crop by addressing these commonly asked questions and typical challenges.

www.ingramcontent.com/pod-product-compliance
Lightning Source LLC
Chambersburg PA
CBHW070213230526
45471CB00002B/938